MW01343737

WHAT IS ORNITHOLOGY?

Ornithology is the study of birds. It explores birds' traits, **habitats**, biology, and interactions with the **environment**, as well as what can be done to protect them.

The scientists who study birds are called **ORNITHOLOGISTS.**

Words that are tricky to understand are in **bold**. Find out what they mean in the glossary.

Words that are difficult to say are in *italics*. Find out how to say them at the back of the book.

ARE BIRDS MODERN-DAY DINOSAURS?

DISCOVER THE SCIENCE BEHIND **ORNITHOLOGY**
(or-nih-THO-luh-jee)

Written by Olivia Watson
Illustrated by Valeria Abatzoglu

More than 150 years ago, a farmer in Germany found a strange **fossil** that looked like a mixture of a bird and a dinosaur. Unsure of what it was, he sold the fossil to buy a cow…

Little did the farmer know, this fossil would completely change scientists' thoughts about where birds come from!

The fossil was studied by fossil scientists who named it *Archaeopteryx*. They worked out that it lived millions of years ago, when dinosaurs ruled the Earth. But the Archaeopteryx seemed very different from huge **sauropods** and **theropods**...

Archaeopteryx was small – no bigger than a pigeon – and looked like it had been covered in feathers. It also had wings, which suggested it might have flown through the sky like so many birds do today. It was very much like a bird, but that wasn't the end of the story!

VELOCIRAPTOR TEETH
VELOCIRAPTOR FINGER CLAW
VELOCIRAPTOR TOE CLAW

After lots of studying, scientists discovered that Archaeopteryx was a relative of the theropods, such as *Velociraptor*. They both had teeth, three claws on their toes, and three fingers ending with claws. Archaeopteryx really was a mixture of bird and dinosaur! This proved the earliest birds **came from dinosaurs.**

It also helped scientists discover that lots of dinosaurs had feathers!

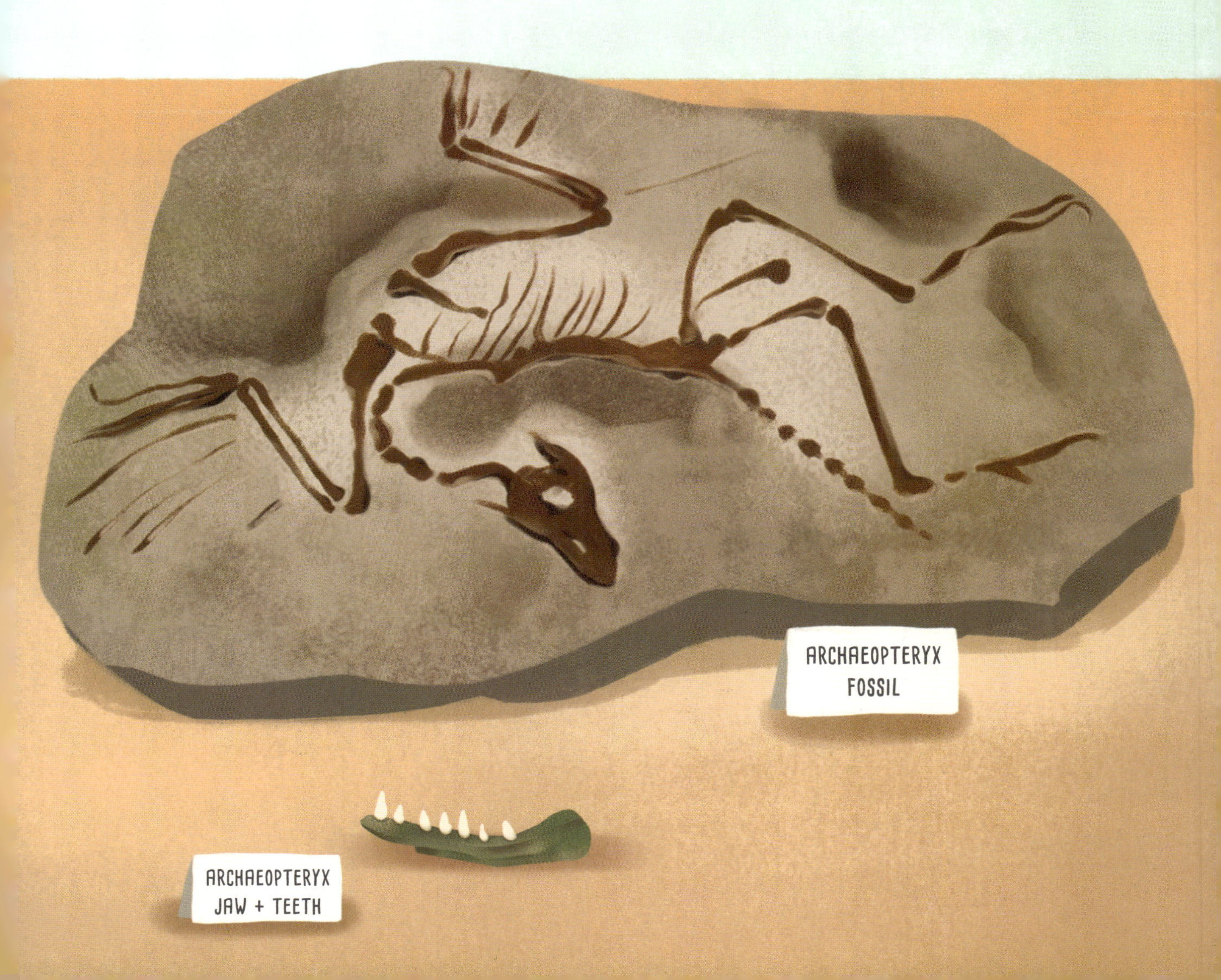

So, what happened to the dinosaurs? About 66 million years ago, a giant **asteroid** hit the Earth, wiping out nearly all of the animals that existed. The mighty dinosaurs that had ruled the planet for millions of years were wiped out.

Or were they...

Small **mammals**, fish, and **reptiles** survived the mass **extinction**. This included a small relative of the dinosaurs – toothless birds, like the *Tsidiiyazhi abini*, which lived around 62 million years ago. The world was a harsh place to live back then, but scientists think these birds survived because they were small, used their beaks to eat small plants, and could fly.

But not all birds were small! Some **prehistoric** birds grew to incredible sizes. Terror birds, which ruled Earth for millions of years, were so huge that they could have put up a fight against the fiercest **predators**!

Luckily, these fearsome, but flightless, birds are no longer around.

Today, there are more than 11,000 different bird **species** in the world! Scientists called *ornithologists* study and protect the birds living on our planet today. They follow birds across the world, even to wild natural areas, such as the wetlands of Africa where beautiful flamingos live.

Ornithologists have taught us just how important birds are! Without birds, many plants would find it harder to grow. Birds collect seeds by eating fruit, then spread the seeds around **through their poop!**

Sometimes, birds drop fruit
or knock it off branches accidentally.
Where fruit and their seed-filled poop
falls, new plants can grow.

Birds aren't just helpful for nature – some help humans too! Greater honeyguide birds in Africa have a nose for honey. And they are happy to share it with us! They help honey hunters find this natural wonder in the wild by leading them to beehives.

However, not all birds eat such nice things. Just like some of their dinosaur **ancestors**, vultures are **scavengers** that eat dead animals! This may seem gross, but it's an important job that makes vultures the ultimate cleanup crew. Without them, illnesses would spread and the natural world would be less healthy.

Birds have lots of unusual ways of helping keep our planet happy and healthy. They eat **parasites** directly from grazing animals, keeping them clean...

they are natural **exterminators,** eating insects to rescue crops...

and they even spread **nutrients** into the oceans through their poop, which keeps these habitats healthy and full of life.

Thanks to clever scientists, and fossils like the one the farmer found, we know that birds are modern-day dinosaurs, as well as amazing creatures that do so much for our natural world.

With some looking so similar to their prehistoric ancestors, it's clear that birds truly are **today's dinosaurs!**

Velociraptor
Golden eagle

Incredible

BIRDS OF THE WORLD

There are thousands of species of birds in the world today. They come in all different shapes and sizes, and each one has special features. Here's a tiny selection of some truly magnificent birds.

Small but mighty BEE HUMMINGBIRDS!

The smallest birds in the world, bee hummingbirds can easily be mistaken for bees! They can flap their wings up to 80 times per second. This makes the humming sound that gave them their name.

Racing OSTRICHES!

Not only are ostriches the biggest birds in the world, they're also the fastest running birds! But they're not able to fly.

Beautiful FLAMINGOS!

Flamingos aren't always pink! These birds are born with white feathers which become pink over time, thanks to all the shrimp and algae that they eat.

Speedy PEREGRINE FALCONS!

These impressive birds are the fastest animals in the world. They fly faster than cheetahs can run! Strong and powerful predators, they are at the top of the food chain.

Unusual POTOO BIRDS!

These strange-looking birds are nocturnal – they sleep by day and hunt by night. With huge eyes for spotting food, and huge mouths for catching it, they are excellent night-time hunters.

Peck-tacular

BIRD FACTS

Birds are incredible animals that have been around for millions of years. We have seen a few ways that they help nature, but what else do these clever creatures do?

FEATHERY DENTISTS!

Some brave birds climb inside the mouths of big animals, like hippos, and clean their teeth by eating food that has got stuck! This helps both animals stay healthy.

MASTERS OF COMMUNICATION

Bird calls don't just help other birds. Scientists have found that woodland animals like birds and squirrels use sound to warn each other of approaching predators!

KEEPING NATURE BALANCED

By eating lots of insects, birds stop too many plants in an area being destroyed. This means that the plants can keep the soil and air healthy, and plant-eating animals still have plenty to eat too.

SPREADING SEEDS

Birds help spread seeds in many ways, including by flight. Some seeds can stick to a bird's feathers and be carried long distances before falling off. Eventually, these seeds may grow into new plants.

SPREADING HAPPINESS

Birds don't just spread seeds, they spread happiness too! Many people find that watching birds or listening to birdsong relaxes them and makes them happy.

GLOSSARY

Ancestors – living things that came before in a family line. For example, your parents and grandparents are your ancestors.

Asteroid – a space rock.

Environment – everything that is around us.

Exterminators – something that completely gets rid of another thing.

Extinction – when something, like a plant or animal species (see right), no longer exists.

Fossil – the remains of a plant or animal that lived long ago.

Habitats – the places where animals or plants live.

Mammals – warm-blooded animals (including humans) that produce milk to feed their young.

Nutrients – substances or ingredients that plants and animals need to live and grow.

Parasites – an organism that lives off another living thing.

Predators – animals that hunt other animals for food.

Prehistoric – the time before humans existed.

Reptiles – a group of cold-blooded animals, including snakes, lizards, and dinosaurs.

Sauropods – a type of huge dinosaur that had a long neck and tail, trunk-like legs, but a small head.

Scavengers – animals that eat animals that are already dead instead of hunting for living animals to kill and eat.

Species – a group of living things that share characteristics and features. For example, greater and lesser flamingos are different species.

Theropods – a type of meat-eating dinosaur that walked on two legs and had short arms. Velociraptor and T.rex were theropods.

HOW DO I SAY?

Archaeopteryx
AR-kee-op-TE-rix

Ornithologists
or-nih-THO-luh-jists

Ornithology
or-nih-THO-luh-jee

Tsidiiyazhi abini
si-dee-YAH-tzee AH-bin-ih

Velociraptor
vel-OH-si-rap-TER

THE BIG QUESTIONS ANSWERED

This is more than just a series of books; it is a complete resource. Accompanying each book is a variety of FREE material to engage curious kids with science.

www.thebigquestionsanswered.com

Use the QR code to visit the website, download free resources, and discover other books in the series.

On the website, find out incredible things about ornithologists, including what they do, some of their greatest discoveries, and the people who have made a difference in this field of science.

The material is also available for home or classroom use, supporting all the information in this book.

Teachers' & Parents' Resources
With discussion prompts and questions, extra information, and facts around key topics.

Young Ornithologists' Activity Pack
Fun activities for wannabe bird experts, including creative writing, drawing, word searches, and much, much more.

The Big Questions Answered is published by Beetle Books.
Beetle Books is an imprint of Hungry Tomato Ltd.

First published in 2024 by Hungry Tomato Ltd
F15, Old Bakery Studios, Blewetts Wharf, Malpas Road,
Truro, Cornwall, TR1 1QH, UK.

ISBN 9781835691366

Copyright © 2024 Hungry Tomato Ltd

No part of this publication may be reproduced, stored in a retrieval system, or transmitted in any form or by any means, electronic, mechanical, photocopying, recording, or otherwise, without prior written permission of the copyright owner.

A CIP catalog record for this book is available from the British Library.

With thanks to:
Editor: Holly Thornton
Editor: Millie Burdett
Senior Designer: Amy Harvey
The team at Beehive Illustration

Printed and bound in China.

Picture Credits:
(t = top, b = bottom, m = middle, l = left, r = right)
Shutterstock: Al'fred 32mr; EcoPrint 34mr; KensCanning 33ml; Mike Truchon 34bl; muratart 35mr; PeopleImages.com - Yuri A 35bl; Rudmer Zwerver 35tl.